IN CHANAK
WITH THE BRITISH ARMY

A corner of Main Street, Chanak.

IN CHANAK

WITH THE BRITISH ARMY

SOME IMPRESSIONS

By Z.
(P.J. BOTHWELL)

EDITED *by* BERNARD DE BROGLIO

LITTLE GULLY PUBLISHING

2023

All rights reserved. No part of this book may be reproduced in any form by electronic or mechanical means, including information storage and retrieval systems, without permission in writing from the publisher, except by a reviewer who may quote brief passages in a review.

Originally published by S. Dirmikis & Son, English Booksellers, Constantinople [n.d.]

Annotated and illustrated edition, Little Gully Publishing, Mosman, N.S.W., 2023

Introduction, appendices and maps © Bernard de Broglio

ISBN 978-0-6459276-4-1 (paperback)
ISBN 978-0-6459276-3-4 (ebook)

Little Gully Publishing
littlegully.com

A catalogue record for this book is available from the National Library of Australia

CONTENTS

Publisher's note vii

CHANAK 1
CHANAK (VERSES) 8
CHANAK ROADS 10
CHANAK HUMOUR 14
LEAVE 19
KHAKI 21
THE PATRIOT 23
SHOPPING 26
MAIDOS PIER 29
THE STAFF 33
THE REGIMENTAL COMMANDING OFFICER . 36
THE DEPARTMENTS, CORPS AND SERVICES . 39
THE CHAPLAINS 42
THE POST OFFICE 44
"TOMMY" 48
A COMPLIMENT 58

Appendix I: A brief biography of P. J. Bothwell 60
Appendix II: The YMCA at Chanak 78
Appendix III: British forces at Chanak 87

Abbreviations and acronyms 97
Bibliography 98

PUBLISHER'S NOTE

This brief, affectionate picture of the British Army in Turkey was written by a peripatetic Englishman, Percival James Bothwell, under the pseudonym 'Z'. The book was published as a slim paperback by S. Dirmikis & Son, Constantinople, probably in 1923.

'Chanak' is short for Chanak Kale, the archaic rendering in English of the name Çanakkale (pronounced cha-nak-kah-leh), a town on the Anatolian shore of the Dardanelles. Çanakkale (literally 'pottery fort') was once renowned for its ceramics. Today it is a bustling university city and tourism hub. But its importance has always been predicated on its strategic location on the Dardanelles, the storied waterway that separates Europe from Asia and connects the Mediterranean with the Black Sea.

Çanakkale sits astride The Narrows. Here the Dardanelles strait is barely 1,400 yards

(1,300 metres) wide, making it the obvious point upon which to concentrate a defence.

In 1915, Çanakkale found itself at the centre of world events when a British and French fleet attempted to take Constantinople (Istanbul) by blasting their way through the Dardanelles. When the warships failed, the Allied Powers decided to land an army to take the forts from the rear, thus setting in train the tragedy now known as the Gallipoli Campaign.

Seven years later, in 1922, Çanakkale found itself again at the centre of world events when Turkish nationalist forces marched upon the town. These were the last days of the Ottoman Empire, when the Sultan sat impotent in Constantinople, and authority lay with a breakaway government in Ankara led by Mustafa Kemal, a hero of Gallipoli. The Turkish nationalists bridled at the dismemberment of the Ottoman heartland by the Allied Powers, in particular the ceding to Greece of large parts of Thrace and Anatolia.

A Greek army landed at Smyrna (Izmir) in May 1919 and pushed inland. They were halted just 50 miles (80 km) west of Ankara at the Battle of the Sakarya in September 1921. A year later, the Kemalists routed the Greek army in Anatolia and swept towards the Aegean coast, entering the great Levantine city of Smyrna on 9 September 1922.

Mustafa Kemal now looked north to the Dardanelles. It was, as ever, the key to capturing Constantinople. In his way stood Chanak, occupied since November 1918 by the Allied Powers. But with interests elsewhere, France and Italy withdrew their forces. Britain alone determined to stand fast at Chanak. She hastily assembled elements of the army, navy and air force to defend the Dardanelles. On 22 September 1922, Turkish cavalry came into contact with the British outposts. War was but hours away.

On this occasion, diplomacy, backed by arms, won the day. And as negotiations ran their course, the British Army maintained its garrison at Chanak alongside a Turkish civil administration.

In 1923, the Treaty of Lausanne formally concluded the conflict that existed between the Allied Powers and the Ottoman Empire, now

Kemalist cavalry in no-man's-land.

succeeded by the Republic of Turkey (Türkiye). British forces quit Constantinople and Chanak on 4 October 1923.

This book was probably written or completed during that period of relative calm in 1923 when the Union Jack and the Star and Crescent flew side-by-side over the town.

But who was its author, 'Z'?

By chance, the publishers saw a copy of the book at the National Library of Australia. It had been acquired by the library in 1981 as part of the Cyril Sidney Hertz collection. 'Sid' was for many years a clerk in the Australian postal service, then a travel agent. He was also a collector of military history books and ephemera, amassing a large collection over the course of 30 years. It was probably Sid who wrote 'P. J. Bothwell' beside the pseudonym 'Z' on the title page of the copy now kept by the National Library of Australia. With this name, the pieces quickly fell into place.

Percival James Bothwell was an Englishman who joined the British Army as a young man and was posted to India, where he would settle and start a family. During the First World War, he gained a commission with the British Indian Army. He seems to have been a sober man, with a vocation for service. These qualities led him to the Young Men's Christian Association, with whom he served at Chanak, then various posts in Protestant denominations and congregations. Bothwell seems rarely to have stayed in one spot for long, and indeed shifted from one hemisphere to the other in 1927, taking his young family to the Antipodes. They spent time in a multitude of towns across New Zealand and Australia, although Bothwell would live out his life in Sydney. A brief biography of the author is included as an appendix.

It is likely that Sid Hertz acquired his copy of the book directly from Bothwell. Another Australian collector, the military historian John Laffin, also had a copy, which was donated upon his passing to the library of the Royal United Services Institute for Defence and Security Studies in Sydney. Perhaps Laffin acquired his copy in the same way. Interestingly, Bothwell seems not to have mentioned his book in the many lectures he gave on the Near East. Or, if he did, it was not reported in the press.

This edition of Bothwell's book is a faithful reproduction of the original text. The first, and only, edition had just one image — the photograph of Chanak that provides the frontispiece to this book. Further contemporary photographs have been added, as well as two maps. Bothwell reported from Turkey for the YMCA's magazine, *The Red Triangle*, and these articles have been included, as well as an indicative list of British forces at Chanak.

Bernard de Broglio

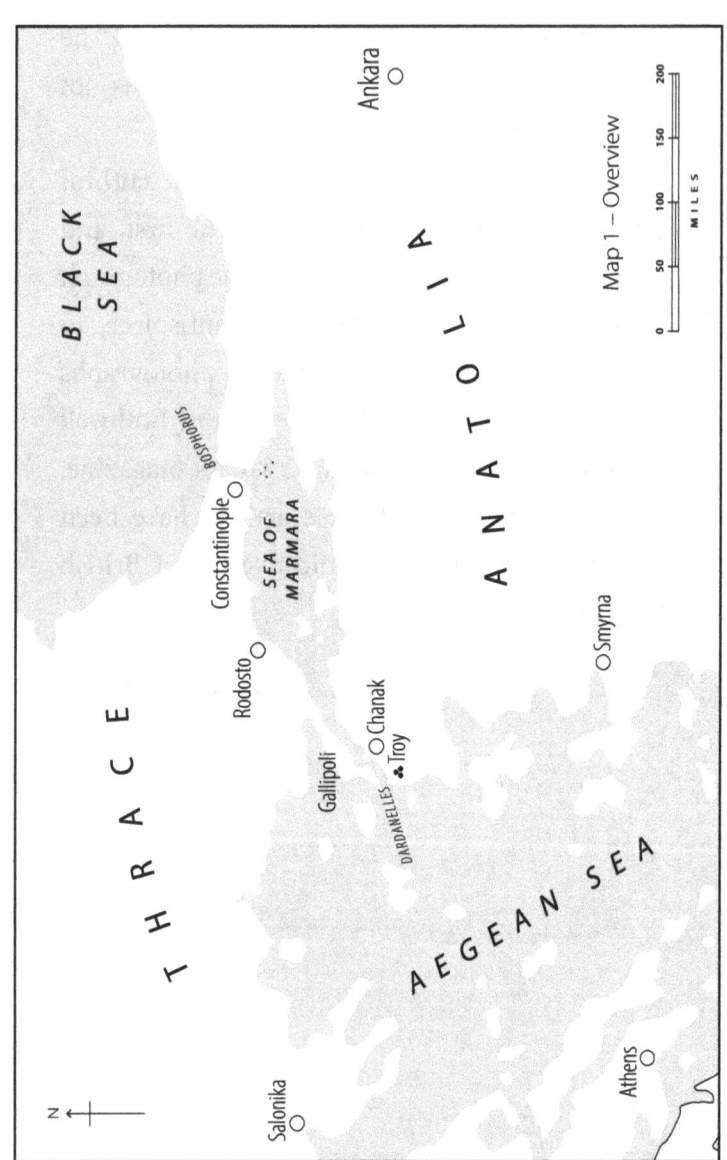

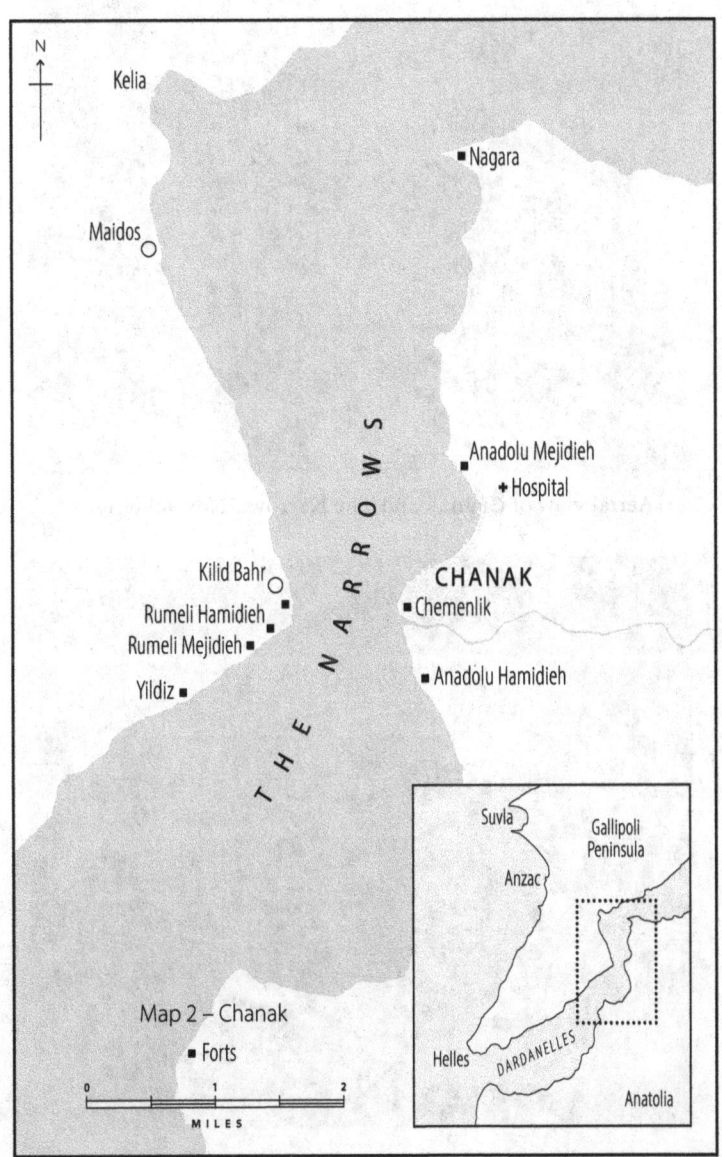

Aerial view of Chanak and The Narrows, November 1922.

Looking south over Chanak on 23 December 1922.
The road to Erenkui and Ezinè can be seen, as well as
Hamidieh Fort on the Dardanelles strait.

CHANAK

WHATEVER one may unflatteringly feel with regard to Chanak as a place of domicile, it has to be admitted that it is a vastly important town from a military point of view and that its strategical position well entitles it to the name of "The Key to the Narrows" (that strip of water, just a mile wide, which constitutes the gateway to Constantinople and the Black Sea *via* Gallipoli Strait and the Sea of Marmora). One perhaps needs actually to be in the place and to stand on the ramparts of the forts of Hamidieh, Chemenlik, Mejidieh or Nagara to realise its possibilities as a stronghold and as a defence against all maritime progress through the Dardanelles beyond it — possibilities that were fully taken advantage of by the Turks who held Chanak during the Great War. It is then not difficult to understand its impregnability from the sea and how it acquired fame (or notoriety, according to whether from the Turks' or Britain's and her Allies' point of view!) in successfully resisting all efforts on the part of British and

French battleships — and the efforts were magnificent, such incalculable results depending upon success — to get beyond, or even up to, the town, especially with the Turks in possession of the Gallipoli peninsular opposite, also well fortified. A truly wonderful exploit, however, was that of a British submarine which, daring gunfire and mines, penetrated into the open sea, reached Constantinople waters safely and returned, executing much damage to enemy ships *en route.*

Today matters are reversed, so far as the occupation of Chanak is concerned, for now the British hold and defend the town; yet not against an expected attack from the sea but from a possible one on land, the line of defence being well-built trenches facing the country and hills in the rear. There is also another difference, it must be noted: the Turks against whom the town is now defended are not those of earlier days, but the Young or Nationalist Turk, more familiarly known as Kemalist. Another sign of the reversal of the former state of affairs is that British warships of all kinds now anchor in Chanak waters or pass to and fro unhindered. Chanak, of course, came into the possession of the British at the end of the Great War.

At the present time Chanak is in the hands of the British alone, so far as military occupation goes. For how much longer is a matter for Allied Statesmen to decide (not unaided by Turkey herself), but it is the devout wish of all who sojourn there that their departure will be in the near, and not distant, future. For Chanak, with all its history and military importance is not one of the earth's most glorious spots.

The town is very dilapidated-looking with but few good buildings and none of them handsome in external appearance: moreover the place possesses a number of shattered and ruined houses, which, presumably, were bombed or shelled during the Great War. The average dwelling-place is jerry-built (as our troops billetted in it have discovered!) : the houses are irregular in style, some possessing rather strangely-constructed interiors, and almost without exception are dirty in their external colouring. This is in the best quarters; in some of the back tortuous streets there is much congestion and the dwellings are mere hovels. From all of which one recognises the typical Turkish town. It is one of those towns of the Near East which never seem clean or attractive in appearance, despite the good work of the sanitary section of the medical authorities on the streets and roads: the general aspect of the place naturally goes against it — photographic views notwithstanding! And in wet weather its roads are vile! But of this more anon.

The present position of the British in Chanak is not without its piquancy, for there also is within the town a Kemalist civil governor (who arrived in December 1922), together with a small force of Kemalist *gendarmerie*. Their duties, naturally, are entirely civil and concern the administration of the law as applied under Kemalist *régime* to Ottoman subjects not in British military employ. The Union Jack and the flag with the White Star and Crescent therefore fly over the town together!

A street in Chanak.

CHANAK

AS VIEWED BY THE BRITISH SOLDIER

THIS aint no blooming city grand,
With nice paved streets or London *Strand*,
And nightly jazzes with a band:
>It's just Chanak.

There aint no women here to cheer,
No chance of courting maidens dear;
There even aint no kiddies here!
>It's just Chanak.

Our billets' not in houses fair,
With loverly carpets on the stair;
We sleep upon a floor that's bare,
>In old Chanak.

There aint no posh electric light
To show us to our *kips* at night;
By candle-light we read and write,
>In old Chanak.

It's just a blighted Turkish town
(O'er which the British flag is flown),
Where, sitting tight, we wait — and moan,
    Is old Chanak.

San fairy an, we aint dead yet
(And don't intend to be, you bet!),
And one fine day a ship we'll get
    'Way from Chanak!

P. J. B.

(Contained in a Christmas Card issued by the Y.M.C.A.)

* * * *

CHANAK ROADS

By courtesy the roads are called such in Chanak; they are really ways or approaches, for except in a few cases they are unmetalled, bumpy and holey. Their condition in wet weather — and it is in the wet season that these lines are being penned — almost beggars description, for in the absence of a drainage system the water stands where it has fallen for days on end. They would make the old road-makers of Britain — to wit, the Romans — weep. There is more than one class of man in Chanak to-day who nearly cries over them: they are the R.A.S.C. motor-drivers, the Signal Service cyclist Despatch Riders, and the R.A. and *Kulmak* drivers of horsed vehicles.

The *pukka* roads or streets are very few, and even they are unevenly cobbled with stones (which, with their bumpiness, are no less the despair of drivers of motor and horse vehicles than are the ill-made roads of the other type); occasionally, too, there is a pavement — of a kind.

The absence of street lamps makes walking out on a dark night when no moon is visible a rather precarious matter, especially when water and mud are feet deep; the convenient pocket-torch becomes a boon and a blessing to men and is rather an ubiquitous article at night

From the point of view of rural landscape there are no delightful walks to be had in the day-time, either. There are no countryside lanes which insistently remind one of England, nor flower-gardens to charm the eye. The redeeming point about the long road to Nagara is the charming sea view which fronts it with numerous and various naval and other vessels riding at sea, forming especially on a fine day a picturesque scene. In the other direction, on the road to Erenkui and Ezinè, past Hamidieh Fort and the control-post, the range of hills whilst certainly green are not majestic enough to be inspiring and the intervening country is flat and uninteresting, whilst, on the other hand, the sea is usually devoid of the shipping which makes the walk on

Gordon Road.

the Nagara road pleasing; it is an excellent road, however, for a long brisk walk of exercise, for a ride, or for training in cross-country running. That is, when dry!

Thoroughfares of any size in Chanak have been given Numbers by the British authorities as "streets" or "avenues" as marks of identification. Some regiments have attempted to immortalise their stay in the place by naming certain of the streets they have occupied with the name of the regiment, such as Gordon Road (the 1st Gordon Highlanders), Sherwood Avenue (the 2nd Sherwood Foresters), Loyal Street (the 1st Loyals), Pack Street (the 14th Pack Battery). One feels that it is a pity that these names will cease to exist when the British occupation has ended, although the units concerned are not in danger of suffering in dignity nor in reputation, the streets or roads being such as they are! The road leading up to the Nightingale Hospital (now used as a barrack, a more suitable building being used as a hospital) has been named Florence Road as a memento to Florence Nightingale.

CHANAK HUMOUR

Old Soldier to One of the Last Draft: "Have you seen 'Billy', the goat, our regimental mascot?"

Draftee: "Aye."

"Well, he's going to be shot!"

"Why?"

"For *acting the goat* in front of yon officers' mess!"

(Quick disappearance of old soldier.)

* * * *

It was his first experience of rain in Chanak and of Chanak roads at such a time, and as he crossed the road as carefully as he could the mud and slush covered his boots, percolated through the lace-holes and reached his socks. On arriving at what passes for the pavement a man of his regiment passed by; *he* had seen quite a few wet days in Chanak.

"Gosh!" said the new-comer, "aint there some mud here!"

Said the other with a superior air. "Mud? Where — what mud?"

* * * *

In full view of the regimental sergeant-major the man passed an officer without saluting. With a commendable effort of self-control the R.S.M. warded off a fit. "Come here, you! ... Don't you know an officer when you see one? Why didn't you salute...?"

"Oh, that's all right, sir," said the delinquent; "you see, he's my Company officer and this morning he gave me five days' C.B., so we're not on speaking terms now."

Then the R.S.M. did collapse and gentle hands bore him into his billet whilst someone ran for the doctor.

* * * *

Ocean liner RMS *Homeric* passing Chanak on 12 February 1923.

She sailed slowly and majestically through the Narrows into Chanak waters. A large and beautiful vessel, obviously a ship of the first calibre in size and appointments. She was the White Star liner the "Homeric", and as she came to anchor outside Kelia the eyes of all the troops along the sea front were quickly and admiringly focussed upon her. They naturally began to speculate upon the reason of her presence: moreover she seemed devoid of all passengers at that distance. The garrison at the time contained a number of time-expired men, due Home for discharge. Ah, that was it: the unknown ship had come to take them away!

The writer met one of the interested men later in the day, a farrier of the Royal Artillery: he had happened to stand with him as the "Homeric" sailed in during the morning. Said he: "What do you think that ship came for? There were a lot of blinking Yankee tourists on board and the boat was only on a blasted tour round!"

* * * *

The Soldiers' Parson and the Soldiers' Smoke.

We have heard of "Woodbine Willy", *
A fearless Chaplain, preacher bold,
But the tale of "Woodbine Tommy"
Remains yet to be told.

Tommy and eternal "Woodbines!"
You cannot part him from this smoke,
Nor tempt at all with other lines:
It's "Woodbines" till he's broke!

(In Chanak as well as elsewhere.)

* * * *

* The Revd. G.A. Studdert Kennedy, M.C. Chaplain to the Forces.

LEAVE

LEAVE, usually of a week's duration, to Constantinople, granted to both officers and men whenever circumstances permit, is a feature of military life in Chanak. Constantinople is some 120 miles distant and the average steamer does the trip in about twelve hours. The leave is regarded somewhat similarly as was leave from France to England during the Great War: at any rate it is much looked forward to. Constantinople is not England, true, but it is a big and interesting city and provides a real change from Chanak: it probably is the most cosmopolitan city in the world, and there is much about its life, especially in the Pera quarter, that is very "continental." And the *lira* does not go far there! But whilst Constantinople may hold much gaiety and laughter, it also holds much sorrow and many tears (a "city of laughter and tears" indeed), for there you meet the poverty-stricken refugee — Greek, Armenian, and Russian (the Russian of high birth in particular) — begging a livelihood. Wherein Chanak, with all its unattractiveness and simple life, perhaps scores.

KHAKI

The most predominant colour in clothing in Chanak is khaki, which is natural. The uniform for the most part is drab almost to the point of ugliness, clean as it may be: there is nothing striking or attractive about it (the officers certainly have a smarter pattern and wear more to relieve its appearance). But it bespeaks serviceableness and efficiency, as becomes the most efficient army in the world for service. It is a khaki army throughout, from General to Interpreter, and of all arms, departments and services as revealed by cap-badge or shoulder-title, and the very simplicity and plainness of the attire indicate the army's thoroughness and readiness for all work, the rough and the smooth. It is a *service* dress and as such possesses its own inspiration to do and dare.

Artillerymen at Chanak.

THE PATRIOT

A SMALL group of soldiers were talking together one evening, or, rather, two were doing the talking: the talking, again, was really arguing. The two were Lancashire lads and one was contending for the town of his birth.

"Aye," said he, "Bolton's th' place! Bolton, the pride of Lancashire — 'tis for sure!"

"Away wi' thee!" was the retort; "how aboot Preston, and Wigan, aye, *and* Manchester? Bolton? Why they're all sae doon-at-heel and puir there that thou hadst to join th' Army!"

"Thou don't know what thou'rt talking aboot, lad! Aw'll tell thee why Aw enlisted. See yon flag there — th' Union Jack? *That's* why Aw joined oop!"

There was a general laugh at this, a laugh more than merely tinged with scepticism, it must be confessed. But the Bolton lad might have been right, after all. The Flag, the life of a soldier, the trappings and accompaniments of martial glory, still have their grip upon, and fascination for, England's young men (a matter for England's deep gratification), and men still "join up" for what soldiering offers more than for what it may provide an escape from. And they are the true patriots of their country.

The Royal Sussex digging trenches.

Outposts on the Chanak perimeter.

SHOPPING

CHANAK is a great place for shops. In the principal thoroughfares especially almost every second door leads into one. And they all are William Whiteleys, Universal Providers — on a scale — for they contain a wonderful conglomeration of articles and stores, from boot-laces to mouth-organs, pills to safety-razors, socks to electric torches, collar-studs to condensed milk, picture post-cards to tinned provisions. At least at times all these articles are to be found in the one shop although it is remarkable how often an enquiry for the very thing one wants meets with: "Ah, Messieur — finished!" (This is a stock phrase known to all shop-keepers!) And the things the shopman sells are all at a price — his own. Still, one has to remember the Turkish customs tariff.

Chanak appears to abound with cafés and *brasseries* (eating and drinking places), of a fourth rate order and yet with high-sounding English designations. As a matter of fact nearly every shop of any size has a corner set apart for the dispensing of tea or coffee or other drinks: but both the tea and coffee as prepared à la *Turque* require a cultivated palate to appreciate their flavour.

Next to cafés, etc. perhaps barbers' shops come in numerical order. And a class of store we are thankful to have is the vegetable one. The crockery and iron-ware shops in No. 13 Street supply many wants to officers' and other Messes.

Since the departure of the Greeks and Armenians practically all the shops are managed by Turks and Jews, although there still are many shops closed and shuttered to be seen, their owners having suddenly left, and they present a rather pathetic sight.

There are two institutions purely British in the town, probably the only two that are so, which specialise entirely in meeting the needs of British troops by the supply of articles of English manufacture and of tea made in the old English way. They are the Navy, Army & Air Force Institute (familiarly known as the "Naafi") and the Army Young Men's Christian Association, and they unquestionably do meet the needs of the soldier, if the patronage accorded the premises of these concerns count for anything. The N.A.A.F.I. have branches all over the town, including canteens "wet" and "dry," refreshment-institutes, and "officers' shops" (a great blessing), and they deal in a very extensive range of commodities. Incidentally they occupy some of the best buildings Chanak possesses. The Y.M.C.A., on the other hand, although, as is their wont, in addition to their dry canteen work carry out a programme of social and religious activities, have only two branches and (rather strangely) are poorly housed at that.

Yes, for obtaining a deal of what one wants along the line of material needs there are many worse places than Chanak.

MAIDOS PIER

(LATER KNOWN AS "CANTEEN PIER")

I WONDER how often the stolid British Garrison Military Policeman on duty at Maidos Pier, in witnessing the departure of parties of Greek and Armenian refugees from that spot, realised that he was a spectator to a series of human tragedies?

* * * *

A small country *droski* was wending its way along one of the less frequented streets of Chanak; with it were a man, a woman and a child. On the cart were all their worldly possessions — a few chairs, a table, a couple of packing-cases, some unpacked domestic articles and a sewing-machine. As the rickety conveyance jolted along, a piece of crude Chanak pottery fell off: the woman gave a cry as it broke on the ground. It was not of any intrinsic value but it possessed a sentimental one and there was little enough of the old home on the cart as it was. The party trudging along at last came in sight of Maidos Pier, where could be seen a crowd

Chemenlik Fort and pier.

of other families, their belongings piled up around them, more than one woman hugging to herself an especially treasured article to ensure its safety. They, too, were leaving the domicile that had been theirs for many years and of which they had grown fond. The landing-place was a babel of voices as the *caiques* or boats to take them away were brought alongside — men shouting directions, women shrilly repeating them, children crying and hanging on to their mothers' skirts.

* * * *

From September 1922 continuously for some weeks these departures took place. The fall of Smyrna with its threat by the victorious Turks to march on Chanak, and, when that fear subsided, the arrival of a Kemalist *governor* and his *entourage* to rule the civil element of the town, filled Greeks and Armenians with an alarm that found relief only in flight to new pastures and to an altogether new life on the Gallipoli peninsular or one of the neighbouring islands. At the same time, too, with the increased British occupation which events made necessary there was a natural demand for increased accommodation of all kinds, and requisitioning of property unfortunately had to come into force — one of the hard penalties of a military occupation.

Maidos Pier was the last spot of their old Chanak that these refugees touched. And, as I say, I wonder how often the G.M.P. on duty there realised all the poignancy of these departures, of what they meant to the individuals concerned. To what were these people going and how have they fared?

A street on the waterfront.

THE STAFF

In a book written in a somewhat light vein on Constantinople I read recently that at Tersane there is the office of a certain civic body which is a "sanctum where the big people sit in solemn conclave, and do it well!" This remark if applied to the staff officers of the Chanak area would be a libel, for they do something more than merely sit. The evidence? It is seen in the fact that throughout all the days of the past months (and some of the days, we learn afterwards, have been fairly critical) the town of Chanak has experienced calm and tranquillity. Plans, counter-plans, provisional plans, for immediate execution if necessary, have been worked out — by The Staff. They have always had a scheme to meet any and every situation that might arise. They do not tell us so openly but somehow we in the place have known of it and consequently have felt, and feel, secure and safe. The slogan of The Staff is: "Preparedness. Never Be Caught Napping;" and the state of preparedness which has existed and exists is such that Chanak is kept free from panic or alarm.

They are an interesting lot, the staff officers. They wear variously-coloured armbands bearing mystic initials which, to the initiated, denote the responsible niche in the scheme of things they occupy. Their responsibilities, though, seem to rest lightly upon them — which is but typically calm British. They are The Staff, of course, and yet if you can get up their sleeve past that red or blue or green (or whatever colour it is) band, and get close to their heart, you find them very human (and you should see them in their Mess!). And now and again they make mistakes (ask the regimental C.O.!), but only the mistakes that really do not matter; in their policies, in the fundamentals, they are "all there," witty, resourceful, and with a solution to every difficulty that presents itself.

I often wonder why no one has nicknamed the staff officers — or at any rate those at General, Divisional or Brigade headquarters, wherever they are situated — the Oyster Bed. For the staff officer for the most part is a regular oyster: he will not open his mouth for you and satisfy your dying curiosity, be your enquiries ever so diplomatically and skilfully presented. If he does it is only to emit a noise like a Cabinet Minister, delightfully vague in its import and beautifully non-committal in its nature, albeit tactfully done. He knows lots of exciting and interesting things days before you are privileged to hear of them, if you hear of them at all: he just keeps them all to himself and his own exalted circle of confrères.

Yet The Staff have to work silently, in which they emulate the wise principles of their Silent Sister Service. They cannot blaze abroad their schemes (brilliant though they may be), nor even their successes (no matter how gratifying). The procedure would be fatal, even if interesting to some people. They cannot, *à la* cock on the dunghill, surmount their oyster-beds and proclaim to the curious world what they have done or are doing. But theirs is the satisfaction of knowing that good staff work has its own reward in what it accomplishes, even if it does not bring a mention in despatches or another ribbon to the coat.

THE
REGIMENTAL
COMMANDING OFFICER

THERE are quite a few regimental C.O.s in Chanak and you can see them every day of your life. In their men.

The drill element in them is to be seen on many occasions — on guard-mounting, sentry-go, company and battalion parades (when not on duty in the trenches), and once a week in real ceremonial style, on Sunday mornings on Church Parade. And the way in which they march and handle their arms, even under the active-service conditions of training which prevail, is a picture for the eye. (You wonder what is the "life" of a pair of boots with all the clicking of heels and stamping of feet that form part of so many drill motions!) The various regiments have their "regimental quiffs," each doing the same thing in a rather different manner, but without exception they are indeed smart, whether it be in the quick step of the Rifle Brigade or the slow measured one of the Highlanders.

You see the regimental C.Os. in the trenches, too — in the science of his construction of the "Front Line," in the patrolling, the work of the look-out men, in the Lewis Gun and trench-mortar sections, in the signallers, and, further back, in the skilful selection of gun positions and the work of the gallant Royal Artillery. The efficiency of all you see, the perfection of the details, fill you with amazement — and admiration.

Then you see him in the lectures in the billets on fire-control, tactics, strategy, etc. etc., and you realise that the Army is a great institute of learning and that the regimental C.O. is not only a clever leader of soldiers but a schoolmaster.

If you know him personally and intimately you discover that he is mighty zealous for the good name and honour of his unit. (His men declare, in fact, that as a delicate compliment to them, he offers them for any and every duty going!) He therefore naturally has an aversion to the things which discount against his unit's good name and honour, principally the "charge sheets" and sick reports which on some mornings appear on his office table. But perhaps his chief aversion is one he does not talk about a great deal: it is the string of red tape which like a telephone wire trails from the Brigade Staff Office to his orderly-room! He may stumble over it occasionally and in any case there are times when its very presence exasperates him. Red tape, however, has its usefulness, and the C.O. is really one of the first to admit it; it ties the various units of the Army up into one efficient and manageable whole. The fact of course is that the regimental C.O. cannot do without The Staff, with all their routine methods, any more than they can do without him: each counts for the efficiency of the other.

THE DEPARTMENTS, CORPS AND SERVICES

"No man can live unto himself"—nor can any one unit of the Army live independently of itself. The fighting unit, perfectly trained in its own line though it may be, cannot live without depending upon one or more other units. The fighting power of a nation consists not alone in Infantry, but also in Artillery; and not only in Infantry and Artillery together, but also in the machines of the Air Force; and not only in these three, but also in the Navy. The fighting power, again, is only possible of functioning when supported and maintained by various Departments, Corps and Services. And this, perhaps, is where the value of the strings of red tape, tying the various Arms of the Service into one efficient whole, is realised. No unit can do without food supplies—hence the presence in Chanak of the Royal Army Service Corps, which Corps also provides fuel and fodder. No fighting unit can do without ammunition,

without tentage and ordnance stores — hence the presence of the Royal Army Ordnance Corps. No unit, whether a fighting one or not, can do without attention to the sick nor without measures for the prevention of disease — hence the presence of the nursing staff and sanitary sections of the Royal Army Medical Corps, and, for horses and mules, of the Royal Army Veterinary Corps. Units cannot manage without telephones, telegraphs, and motor despatch-riders, which explains the presence of the Royal Corps of Signals. And so on, *ad infinitum*. Each of these again is dependent upon one another for assistance, if

The base at Kelia.

not for very existence, in the way of men, food, transport, material.

Space does not permit in this small book to deal in detail with the work of the various Departments, Corps and Services represented in Chanak. Reference has been made on another page to the perfection of Chanak's line of land defence, centred in the regimental commanding officer, and it may suffice to say that this line could not be so perfect nor any C.O. have the efficiency that is his were it not for the work of those who in their depots, bases and workshops "carry on," for the most part unseen but always potently realised.

THE CHAPLAINS

THE British Army in Chanak has been blessed with padres who besides being popular are keen sportsmen and social workers. They are exponents of a Christianity practical as well as theoretical.

The Sports Secretary of the Area at the beginning was a Church of England chaplain and on his departure the Senior Protestant Chaplain (a Non-conformist) assumed the duties, he himself proving no mean Boxing and Football referee. Another Church of England chaplain is the chairman of the Cross-Country Running committee. The Senior Roman Catholic chaplain has been responsible for the creation of a first class concert party, named *The Sentinels*, which in addition to delighting audiences at the Divisional Headquarters theatre dared to pay a visit to Constantinople and showed concert managers there how to put on a programme!

There is nothing namby-pamby about the remaining two chaplains, either: and all are soldiers' parsons, knowing their men and understanding them thoroughly. The 2nd Highland Light Infantry, by the way, have a minister all to themselves—perhaps a necessary precaution, since it takes a Scotsman to understand a fellow Scot's tongue! You have no doubt as to the nationality of the H.L.I.'s padre when you hear him speak!

The chaplains do their share of "theorising," too, chiefly on Sundays. Sunday is indeed their busy day, with parade-services, voluntary services and meetings, from early in the morning till late in the evening. Then during the week there are occasional meetings or classes to conduct, visits to the hospital to make, visitors to receive and counsel.

One chaplain is an expert horseman. Another gave up smoking for a month to prove to an officer that it could be done! And all, although divines, are intensely and sympathetically human.

THE POST OFFICE

THE link between Home and the exile is the Post Office, an estimable institution (when reliable!). For some months now Chanak has possessed a branch of the British post-office, more or less complete in service, the main exceptions being savings-bank work and delivery direct to addresses of their postal articles. It boasts a real live civilian post-master. It was not always so; there was a time when the "post-office," managed by a small but stalwart squad of soldiers, comprised merely a receiving and delivery office for letters and parcels: it did not "register" articles, sell postal-orders, nor even sell stamps! There were many days when stamps could not be obtained in Chanak for love nor money, and the only place where they could be obtained *when in stock* was the N.A.A.F.I. canteen. This undoubtedly was a hardship to the troops generally. Orders regarding the franking of unstamped letters by commanding officers — *vide* King's Regulations — did not issue for some time, and the War Office had ruled that as the troops in Chanak could not be considered as on "active service" (a really anomalous ruling, considering that for certain other purposes the men were on active service) free postage was not permissible. Meanwhile hundreds of letters, unstamped, marked "On Active Service" or "No Stamps Available," were posted, only to be charged double for on arrival at the place of address! Which uncommonly looked like penalising the soldier for being in Chanak—as though he had not already enough to put up with in the place!

As already hinted, one has still to attend personally or by servant at the post-office for one's mail. It is wondered, how often it is realised that the obtaining of some of these postal articles, so quietly (and, let it be noted, so courteously) handed over the counter by the soldier or civilian assistant, has involved discomfort in no small degree and even risk to life and limb to some of the staff. Going out to a ship that is only passing Chanak, at any hour of the day or night, often in stormy and wet weather, and getting the mails off by means of a rope ladder (the vessel merely slowing down and not stopping) has proved more than once an uncomfortable and hazardous proceeding on the part of the soldier assistant. It has often had to be done; similarly with the letters, etc. that we send away and of the safe receipt of which we later hear from Home and elsewhere.

In other ways, too, the work of the soldier and civilian assistants at Chanak's B.P.O. does not make for it being the softest of jobs. We feel grateful to the staff, for they give us the letters which mean so much in our exile away from the Homeland. Occasionally delays occur (and they are rather inexplicable), but they may not be their fault, and on the whole the little Post Office as it now exists is a remarkably efficient institution.

An aeroplane mail service between Chanak and Constantinople would be an appreciated arrangement! At present the mails arrive and leave twice weekly, by steamer, with an occasional mail either delivered or taken by a passing ship, and a daily service (such as our fortunate friends in Constantinople have, the continental Orient Express arriving and departing daily) would be appreciated indeed: the aeroplane could effect it.

"TOMMY"

"TOMMY" is a wonderful fellow, in many ways. But he has some queer personal characteristics and because of them he is more or less a puzzle to the average civilian. At times he certainly does show himself to be a bundle of inconsistencies, whims and fancies. In the art of rumour-spreading he is something of a past-master, and it is amazing what he himself will believe.

He seems to forget that he ever was a civilian and thinks because he has donned a uniform common with thousands of others and bears a Number that he has lost his identity as an individual, certainly as a responsible citizen. Which is wrong. He is still a citizen of the Empire, only armed and clothed and living for the needs of war, should war come. He appears to lose sight of this and so we see him doing things as "Tommy" which he would not do as "Mr." It perhaps is a matter for the student of mob psychology to explain. Yet there are occasions when Tommy does recognise himself as a responsible individual, particularly in the performance of his duty, and when he does so the results are not pleasant to those who question his personal authority; and he can when required show both initiative and resource. There was a sergeant on control-post duty one day who demanded from a Kemalist soldier his rifle, to unload it before allowing the man to pass, and on the man at first declining the sergeant promptly took from his own pocket a safety-razor blade and began to cut the rifle-sling! No one but a Tommy would have thought of that. On sentry-go, too, Tommy is possessed with an authority which he never hesitates to exercise.

The confidence he has in himself is nothing short of sublime. Give him a rifle and a bayonet with the order to stop anyone from passing a certain point, and he will stop an Army Corps!

Kipling has written that Tommy is no plaster saint. Nor is he—thank God! If men are to be saints at all let them be alive and full of movement and deed. It is for the making of this kind of saint, surely, that the State provides the Army with places of worship and chaplains. Tommy has a soul all right, and saints in the army are possible, for they

British and Kemalist sentry posts in October 1922.

have been and are; the question is perhaps one of definition of saintliness. There are saints *and* saints! Each man must choose his own ideal. The following lines by Edgar A. Guest is a Tommy's fine prayer:

I would not stand apart nor dwell alone,
    Nor live as one too good to soil my hands;
I would not guard the soul that is my own
    So closely that it shrinks from life's commands
And scorns to go where shame and sorrow reign
    For fear it, too, may wear a scarlet stain

Clean hands at night! That is the pride I ask,
    But let me stand to service through the day;
Let me go gladly to my grimy task,
    I'll bear the dirt which I can wash away.
Though deep in mire Life calls on me to fight,
    What matters that, if I am clean by night?

The chaplains tell us that Christ demands not only belief in Him but service—faith and works together.

I have said that Tommy is a bundle of inconsistencies. One of his most glorious ones is that he will criticise his regiment, calling it a rabble or a mob or a heap, criticise his officers from the C.O. downwards, similarly the N.C.Os., tell you that he is fed up to the back teeth and intends to put in for a transfer, and yet you must all the while remain a silent listener — that is if you are not "one of his." If *you* dare to agree with him, if you in any way say ought against his unit or those in it — well, someone had better go quickly for the provost-sergeant, for there's going to be a fight. Thus is he queerly loyal.

He may be full of complaints, yet let there be an important parade (especially if it's a General's), and Tommy turns himself out for it and drills as if the whole appearance of the regiment and success of the parade depended upon him personally. Let the Inspector of Hygiene announce a visit to the billets, and Tommy enters into the spirit of rubbing and scrubbing and white-washing with tremendous zest and pride, as though he himself were the regiment to take all the praise or censure.

Tommy's propensity for grousing is proverbial. He will grouse over and cuss most of his jobs, but the point is that he gets on with the work.

Which indicates that grousing is nothing more than a habit. It's an inherited habit and is in the blood of the ranks. As is swearing. Tommy will give vent to the most extraordinary epithets and expressions, in even ordinary conversation, but he can't mean them: some of them are so impossible! It would indeed be better if he saved his breath to keep his body warmer (so people say it would also be nicer).

Stick Tommy wherever you like and he will carry out his job there. He may not know a word of the language of the country into which you have stuck him, yet somehow he'll carry on. Be it amongst the Greeks in Rodosto, the Armenians in Anatolia, the Turks in Chanak, the Arabs in Mespot, the Indians in India, the Chinese in China, Tommy gets through his appointed task in some wonderful way, even as he did in the strange lands of France, Belgium, Germany, Italy and Russia.

Tommy's tasks in Chanak have been multifarious. A great deal of the work, particularly at the beginning of the occupation, comprised hard labour, *e.g.* in repairing broken roads and making entirely new ones. As a man was heard to say: "I'll tell you what it is: they're trying to make us a Chinese Labour Corps before we leave!" (His battalion was under orders for China.) But it all has counted for efficiency and for safety, and Tommy has shown that he can do anything and everything required of him, from digging trenches to sick-nursing, from criss-crossing a town with telegraph wires to baking bread, from erecting tin huts to cyclostyling copies of "Orders" and Leafield Press Messages — from, indeed, making towns (witness Kelia) to demolishing them! One thing, though, he has shown a distinct aversion to, and that is his periodical supply of biscuits and bully beef! Of these he considers he has had his full share in Chanak!

Yes, Tommy is a wonderful fellow, his mental and moral complexities notwithstanding. What would we do without him?

Nagara Fort, 1923.

A COMPLIMENT

THE following is a copy of a Special Order by Lieut.-General Sir Charles Harington, G.B.E., K.C.B., D.S.O., commanding the British Forces in Turkey, to his officers and men, issued from Constantinople on the 27th February 1923, after a visit during the previous month to Chanak:

Having just completed my inspection of every unit in the Command, I should like to take this opportunity of expressing my appreciation of the smartness and efficiency which I witnessed throughout.

I was very glad to have the chance of speaking to nearly every man in the Force, and was much impressed by the cheerful way in which everyone has made the best of the local conditions and discomforts.

I am very hopeful that peace will be obtained shortly, but if the reverse should unfortunately be the case, I am confident that no Commander could wish for a more efficient Force than the one I am proud to have under my command.

In it is included a contingent of the Royal Air Force, and behind us lies the powerful Fleet of the Royal Navy ever ready to help us. No finer example could be found of strength founded on friendship than the example of the three Services in the Constantinople and Dardanelles areas, a friendship which none of us will forget.

I have been particularly struck by the efficiency of the Administrative Services and Departments of this Force, and by the care displayed by all Commanders and Staffs in attending to the welfare and comfort of their men.

I take this opportunity of congratulating and thanking all concerned on the result of their labours and for fostering a right good spirit throughout the entire Force.

APPENDIX I

A brief biography of P. J. Bothwell

Percival James Bothwell was born in London on 9 September 1884. He was the first of three children to James (born Edinburgh, about 1856) and Mabel (born Blyford, Suffolk, about 1867). His younger siblings were sister Edith and brother Herbert Ernest.

The family was regularly on the move. In 1891 they were living in Salisbury, Wiltshire. By 1901 they had moved to Dorset where Bothwell's father was employed as a verger. The sixteen-year-old Bothwell was a butcher's clerk.

At 18 years of age, Bothwell enlisted in the 11th Hussars, a cavalry regiment of the British Army. He signed his attestation papers at Dorchester in the county of Dorset, but gave his address as London and his profession as clerk. He was allocated service number 5014.

Bothwell appears to have been absent for much of his short time with the 11th Hussars, and was discharged on payment of 10 pounds less than three months after enlisting.

A few months later, in 1903, Bothwell enlisted with The Rifle Brigade (service no. 9810) and was posted to the 4th Battalion. He was appointed acting corporal in 1904. The following year, Bothwell was granted permission to extend his service with the colours from three to eight years. He had earned one good conduct badge and had completed his certificate of education (second class).

Bothwell embarked for overseas service in November 1905. He was stationed on the island of Malta until December 1906, during which time he was promoted corporal and passed classes of instruction for promotion to sergeant. However he reverted at his own request to rifleman on 27 December 1906 when he was posted to the 2nd Bn, The Rifle Brigade, and began service in India.

Bothwell's service record does not reveal much about his service on the subcontinent over the next four years, except that he was appointed acting corporal (unpaid) in January 1907 and granted another two good conduct badges. Bothwell appears to have been employed as a clerk at various stations, including army headquarters, for much of his time in India. On 23 March 1911, he was transferred to the army reserve on expiration of his period of army service. 'Conduct exemplary,' wrote a senior officer. 'No instance of drunkenness during the whole of his 8 years service.' The 26-year-old Bothwell received permission to reside in India, and he gave his intended place of residence as the YMCA in Calcutta (Kolkata).

The Calcutta YMCA dates to 1857 and was the first to be established in Asia. In 1891, a national association of YMCAs in India was formed. By 1911, six other branches were in operation. Bothwell was secretary of YMCA Bangalore (Bengaluru) by 1913.

That year, on 12 June 1913, Bothwell married Ellen Charlotte Davis Stoddard in St Andrew's

Church, Madras (Chennai). Ellen, born 2 July 1886 in India, was the daughter of a tea planter. The marriage certificate reveals that Ellen was a resident of Simla (Shimla), the fashionable hill station below the Himalayas, and worked in the education department.

At the outbreak of the First World War, Bothwell returned to the colours as Acting Corporal 9810. He was posted to the 4th Bn, The Rifle Brigade, on 26 August 1914.

Less than two months later, on 10 October 1914, Ellen gave birth to their first son, Gordon Harold Stoddard, in Poona (Pune). On 27 January 1915, at three months of age, the child was baptised at St James' Church in Calcutta. The baptism record shows both parents as residents of Delhi, and the father employed as a clerk in army headquarters. Sadly, the boy died before his first birthday.

By now, the 2nd and 4th battalions of the Rifle Brigade had embarked for the Western Front (in September and October 1914, respectively) but Bothwell remained in India.

On 23 March 1916, Bothwell was discharged

from the army on completion of 13 years' service. He signed the discharge at Simla.

That year saw the birth of a second child, this time a daughter, named Pamela Constance.

Bothwell successfully sought an appointment to the Indian Army Reserve of Officers. (This reserve numbered forty-seven men at the outbreak of war in August 1914 but grew to a strength of 4,500 by war's end.) Bothwell was commissioned on 28 October 1916 as a second lieutenant, and attached to the British Indian Army's Supply and Transport Corps. The following year, on 28 October 1917, Bothwell attained the rank of lieutenant. He served in India, perhaps on the North-West Frontier, and in South Persia.

By 1919, Bothwell held the rank of acting captain. He commanded an animal transport company (part of 103rd Pack Mule Corps) from 23 February 1919, and a mule corps from 4 March 1919. (The mule corps provided first-line transport for the army, carrying water, ammunition, entrenching tools, signalling and medical equipment.)

For his service in the First World War, Bothwell

was awarded the British War Medal, Victory Medal and General Service Medal (1918) with S. Persia clasp.

On 18 August 1921 in Murree, another popular hill station in British India (now part of Pakistan), Ellen gave birth to a second son, Percival Roy Bothwell, known as Roy.

In October of that year, Bothwell was initiated into a Masonic lodge in Murree, although he gave his residence as Jhansi in central India. Freemasonry in India was marked by its multiracial character and religious tolerance, an aspect that was presumably attractive to Bothwell. A certificate was issued in March 1922 confirming his initiation into Lodge Light in the Himalayas.

Bothwell resigned his commission in the Indian Army Reserve of Officers on 17 February 1922. Later that year, the thirty-seven-year-old moved his family to England, where Bothwell joined the Toc H movement. He maintained his connection with the YMCA, on whose behalf he was then despatched to Constantinople. Bothwell appears to have spent about a year in Turkey, stationed in the capital and at Chanak, where he was the branch secretary.

While he was at Chanak, Bothwell took the opportunity to visit the battlefields of Gallipoli. In later life he would give illustrated talks that included pictures of the landing beaches and war cemeteries on the peninsula. 'In the course of a few years, Gallipoli will be the scene of great pilgrimages,' said Bothwell in one lecture. 'The beauty of the cemeteries cannot be realised except by those who have been there.'

Bothwell may also have visited Smyrna. In a lecture on the Near East, he is reported as showing a picture of the town burning during its abandonment by the Greeks in 1922.

Bothwell probably left Chanak at about the same time as the British occupying force, which departed in October 1923. This might have been when he visited Malta and photographed Anzac graves in the war cemetery at Pieta, the subject of another lecture he would give in later years.

Bothwell was then appointed Wesleyan

Missionary and acting naval chaplain at Port Said, Egypt. It was in that dry and sandy land that Bothwell decided upon a move to the lush green country of New Zealand. Perhaps visiting the Anzac war graves at Gallipoli and Malta had stirred in him a desire for the Antipodes. On 23 November 1925, Bothwell and his family landed in New Zealand from the steamship SS *Niagara*.

Bothwell took up a pastorship with the Kuripini Methodist Church in Masterton on the North Island. Within a year he had moved to Wellington, where he was inducted into the pastorate of the Newtown Congregational Church. Local newspapers describe an energetic clergyman, delivering talks and lectures, officiating at various ceremonies, including Anzac services, and participating actively in the local community. He was elected president of the Wellington district of the Congregational Young People's Fellowship.

THE CHURCHES.

Rev. P. J. Bothwell.

TOOWOOMBA APPOINTMENT.

The Rev. Percival J. Bothwell, who was for three years pastor of the Willoughby Congregational Church, has left Sydney to take up his new duties at Toowoomba, Queensland.

Mr. Bothwell has had a wide and varied experience. He has lived in no less than eight countries—England, Malta, India, South Persia, Turkey, Egypt, New Zealand, and Australia. He entered the service of the Y.M.C.A. in London and was later sent on service to Constantinople, via Calais, the journey necessitating the crossing of six frontiers. He was engaged in Y.M.C.A. work in India for several years, and during the war was

In late 1927, Bothwell relinquished the pastorate—for 'family reasons,' according to a local newspaper—and moved to Australia. He became pastor of Willoughby Congregational Church in Sydney, was registered as a marriage celebrant in NSW, and later elected president of the Northern Suburbs Congregational Sunday School Teachers' Association. His daughter Pamela studied dressmaking at the technical school in North Sydney.

In 1931, the family relocated to Toowoomba in Queensland, where Bothwell was ordained to the Congregational ministry on 27 February, and filled a temporary vacancy in the parish.

Bothwell and his family threw themselves into local initiatives. These were the years of the Great Depression and Bothwell was secretary of the Ministers' Fraternal soup kitchen committee that provided meals for thousands.

Bothwell's sermons and activities were regularly reported in the local press. He gave 'lantern addresses' on the Near East and Malta, and featured on local radio. Notable is his advocacy for the prevention of cruelty to animals — he campaigned against live hare coursing — and an enlightened attitude to birth control. Bothwell stated in a local newspaper that 'sex-repression is infinitely more harmful than sex-expression.'

> All that I am pleading for is that where the limitation of births obviously is wise for health reasons, or humane for the woman's sake, and just to children already in existence, Christian sanction should be accorded it, and that the prevailing hypocrital (not to say ignorant) attitude towards it should give way to a saner and cleaner outlook upon it and upon sex generally.

In 1933, Bothwell resigned his pastorate of the Toowoomba Congregational Church, apparently due to the church's straitened finances. The reverend, his wife and two children were fondly farewelled by the local community.

Bothwell's next post was the pastorship of East St Kilda Congregational Church in Melbourne, Victoria. His initiative to screen films to increase church attendance and spark debate attracted attention. On one occasion, 500 people attended a church service and film screeening, with 200 more unable to gain admission. Bothwell also advocated for casual dress at church, and suggested amending the hours of worship to give people more time for recreation on Sundays.

In January 1935, Bothwell had a letter printed in *The Herald* of Melbourne on the topic of sex education. Bothwell described himself as an ex-director of a Sydney organisation (the Institute of Family Relations) that aimed to make parents good teachers 'in this subject.'

> The truly extraordinary position is that we deliberately hedge our children in their younger days around with all manner of fairy tales regarding their coming into the world, and then later suffer the humiliation of knowing that they have found our untruths out! Perhaps we ourselves have had to take a part in pulling down the hedge we were so careful to construct!

In 1935, Bothwell accepted the offer of the pastorate of the Bendigo Congregational Church, where he began his ministry in July.

In August 1937, Bothwell made an unsuccessful bid for pre-selection in the Federal seat of Bendigo for the United Australia Party.

By December of that year, Bothwell had resigned his pastorate of Bendigo, and a local newspaper reported him to be in Sydney. However, Bothwell was back in the Victorian town by June 1939, when he was interviewed as the Bendigo organiser of the Boys' Employment Movement. Permanent jobs had been found for half of 1,280 boys who had sought employment through the branch, said Bothwell, who was campaigning for a centre in town where boys could meet and receive elementary instruction in a trade.

Australia entered World War II on 3 September 1939, and in October, Bothwell joined the Citizen Military Forces, the Australian part-time militia, as a Temporary Captain. He appears to have been assigned to the staff of several training camps and depots in the months that followed.

Captain P.J. Bothwell, official military photograph.

Roy Bothwell.

Bothwell's service with the professional military forces dates from 17 June 1940 to 3 March 1943. For some or most of that period he was assigned to sea transport. Bothwell made one return trip (voyage only) to the Middle East, embarking 8 April 1941 and returning to Australia on 15 June 1941. In March 1943, Bothwell was placed on the Retired List in the rank of Captain. His service numbers were V.85858, VX51948 and V145108.

* * * *

Bothwell's son Roy joined the Royal Australian Air Force (RAAF) at age 18 in February 1940. (Roy had previously applied for entry to the Royal Military College Duntroon but was rejected as unfit, 'insufficient chest development.') On graduating from Point Cook in June 1940, Roy was commissioned as a pilot officer, and became one of the youngest officers in the RAAF. He was promoted to the rank of flying officer in December 1940.

Roy Bothwell embarked for the Middle East in February 1941, joining No. 3 Squadron RAAF flying Tomahawks over Syria and North Africa, where he saw several air combats with

Messerschmitt 109s and 110s. In November and December 1941, when the British Eighth Army attempted to relieve the Siege of Tobruk, Roy's squadron fought over the Western Desert as part of Operation Crusader.

On the morning of 25 November 1941, a column of enemy vehicles was spotted heading east through Allied lines south of Sidi Omar. Twelve aircraft from No. 3 Squadron scrambled to strafe the tanks and motor transport. They encountered severe ground fire from the tanks and Roy Bothwell, in Curtiss Tomahawk IIB AM398, crashed into the ground. A compatriot said he flew straight into a tank. The following day, three officers from the squadron went out to recover his body. They thought he might have been wounded before he crashed into the ground. Roy Bothwell is buried in Halfaya Sollum War Cemetery, Egypt, in plot 22, row A, grave 2.

* * * *

After the Second World War, little is known of P. J. Bothwell's activities, except that he was preaching in the Unitarian Church, Cathedral Place, Melbourne in January 1946.

Bothwell died in Sydney on 10 July 1954, aged 69, survived by his wife Ellen and daughter Pamela.

APPENDIX II

The YMCA at Chanak

The following two articles appeared in The Red Triangle, *the monthly magazine of the British National Councils of YMCAs. The first was probably written by P.J. Bothwell, the second is attributed to him.*

AT CHANAK

This place has few attractions for our soldiers. Rows of empty shops, barred and shuttered, tell of the flight of the Greek and Armenian shopkeepers; the absence of little children from the streets, and of women, strikes Tommy's heart with a chill; no wonder he hungers for the life of Constantinople, fourteen hours steamer journey away, and those who had Christmas leave there were envied.

The Association has two centres at work in Chanak, and judging by the patronage of the troops they are meeting a great need. The one in the Eastern section of the town is an old Greek wine-shop, barn-like within, with room for about 200. The other centre is on the main street, in the heart of the town, and is a Turkish theatre with a stage. This, likewise, can put up a couple of hundred. The lack of room and rooms makes it difficult to put up a full programme, but much is being done.

Every night both places are crowded; sometimes in a morning there is a good muster. Whist drives, an occasional concert, tournaments for indoor games, once a week a band concert — all these hold the men. The lending library is much used: more books are wanted.

On Sundays, Army Chaplains conduct services in the evening; there is a weekly Bible Study Class, and enthusiastic members of the Soldiers' Christian Association hold their regular meetings. We held special Christmas services at the two centres; on Boxing Night a pierrot troupe, formed by the men themselves, entertained a packed "house." A special Christmas and New Year's card was issued by the Association, one of the secretaries composing a poem, "Chanak," which was quickly learnt and recited in the "wet" canteen. At least one regiment sent a card home for publication in a newspaper, and though the poetry may not be equal to that of a Poet Laureate, it expressed something of Tommy's feelings about Chanak and the conditions there under which he lives.

Mr. P.J. Bothwell (our secretary) sends a glowing tribute to the British soldier. "Stories of exhibition of wonderful patience," he says, "are current. One is led to admire the qualities of discipline and patience exhibited in the face of acts of irritation during earlier, critical days in the town, when the Kemalist forces were up to (and more than once in) the front-line trenches, declaring that they knew no boundary-line. The confidence he possesses is sublime. I saw, one night in Constantinople, a lance-corporal, single-handed, keep on the move a cosmopolitan crowd, all attracted to a spot still wet with human blood, and the obedience he received was a revelation.

"A look in at either of our branches any night fully demonstrates that the Association's work amongst the troops is worthwhile: it is worth all the labour involved, and the money. Every table is occupied by men reading, writing, playing draughts or cards, or enjoying conviviality with their fellows over a mug of good old English-made tea or cocoa. There they are, grousy over their lot as soldiers in Chanak, but happy in that they have a Y.M.C.A. in which to spend their leisure moments. On a visit to the trenches, the secretary asked the men if they would like a centre, a dry canteen, there for their benefit. 'We would not be half laughing if you came,' a man instantly replied. Thus our presence is appreciated."

TRUST AND SERVICE AT CHANAK

The Y.M.C.A. Secretary is moving amongst the men in his building. A soldier comes up to him with a parcel. "Will you post this for me, sir? It won't go into your letter-box. You know we are 'standing by' and so are confined to the area of our billets, and I cannot go to the post-office myself."

"Yes, certainly. Let me see the parcel ... Well, there are three ways of despatching this. First, by parcel post, in which case you must sign a customs declaration form and affix it to the parcel: I have one I can give you; second, by book post, in which case the ends of the package must be left open, which has not been done; or, third, by letter post, just as it is, except that the postage will cost about four times as much as you have stamped it for. Which way would you prefer it to go?"

"I leave it to you, sir, and here are twenty piasters more to cover extra postage if required."

"What is in it?"

"An album of views."

"All right. I will see it posted for you."

The Y.M.C.A. worker takes the package, as likely as not has to open it and entirely repack it. The amount of postage required is ascertained and the parcel safely posted.

Trust, as well as *Service*.

* * *

Soldier: "Excuse me, sir ... I drew short pay last week; it's a couple of days yet for pay-day, and I'm hard-up and dying for a smoke. Will you let me have a *lira* (value about 2s. 6d.) until pay-day? You can take this ring and keep it until I pay you back."

Y.M.C.A. Secretary: "I will loan you the lira, but I do not want your ring! We are not pawnbrokers!! Just let me have your name and I will leave it to you to repay the money."

The man is profuse in his thanks and proceeds at once to the refreshment counter.

Another case of *Trust* — on the other side!

* * *

Soldier: "Can you get for me some developer?"

Secretary: "Some *what?*" — with an unpleasant vision of Sandow's gymnasium.

Soldier: "Developer — for developing films ... You see, there is only one photograph shop in this benighted town and the photographer will not sell *me* any developing solution, in order to compel me to take my films to him to be developed. I prefer to do my own developing, if only for the reason that it's cheaper. Now, if only one of your native staff could take a bottle to the shop he would have no difficulty in getting some of the solution, and I ... "

Arranged! An instance of *Service*.

* * *

It is snowing heavily and the wind is cold and biting. In the Y.M.C.A. building, not yet opened for the day's work, it is dry and warm. A sergeant gains an entrance.

"Lieut. Brown's compliments: he is orderly officer of the day. Would you kindly allow the guard to 'mount' and the old guard, when relieved, to 'dismount' in your building, on account of the weather?"

The reply of the secretary is, "Of course, by all means!" and for the next half-hour the Y.M.C.A. room is filled by the sound of clanking arms, sharp words of command and much clicking of heels.

Service.

* * *

Y.M.C.A. Secretary: "Hullo, Green. Where's your cap?"

Green: "Lost it. Can't get another: all in the quartermaster's stores are too small for me! My head is not so outrageously big as it is so extraordinary in its length, from back to front!

And needless to remark there are no shops in Chanak which stock soldiers' service-dress caps. So I go about bareheaded, in all weathers and at all times."

"That's unfortunate ... I thought, too, that you had been made a bombardier? You are wearing no stripes?"

"So I have. But, again, there are no chevrons in the stores just now, so I have to go without my badges of rank."

"Well, I can supply your deficiencies, I think. I can ask the Association in Constantinople to send me down a cap and some stripes. They will obtain them if I ask them to."

"I'll be delighted if you will do so! Cap, the largest size obtainable!! I'll fix its shape!"

Done! Cap and stripes received within a week. *Service!*

P. J. Bothwell.

'Old Bill' on outpost duty at Ismid.

APPENDIX III
British forces at Chanak

British Army outpost.

11–26 SEPTEMBER 1922

Dardanelles Sector Defence Force

Officer Commanding, Colonel-Commandant
D. I. Shuttleworth, CBE, DSO

Cavalry

"B" Squadron, 3rd The King's Own Hussars

Infantry

1st Battalion, The Loyal Regiment

1st Battalion, The Gordon Highlanders

2nd Battalion, The Royal Sussex Regiment
(from 18 September)

2nd Battalion, Sherwood Foresters
(from 22 September)

2nd Battalion, The King's Own Scottish Borderers
(from 26 September)

2nd Battalion, Highland Light Infantry
(from 26 September)

Artillery

96th Battery RFA

92nd Battery RFA *(from 13 September)*

19th Brigade RFA, less one battery
(from 22 September)

3rd Medium Howitzer Brigade *(from 25 September)*

Royal Engineers

> One section, 55th Field Company

Royal Navy

> 4th Battle Squadron, including HM ships *Ajax*, *Marlborough*
>
> Shore detachments of more than 1,000 men (Royal Marines, officers and ratings), including 12 to 14 Vickers guns and 24 to 28 Lewis guns with crews attached to infantry battalions
>
> Aircraft carriers:
>
>> HMS *Pegasus* (four Fairey IIID seaplanes)
>>
>> HMS *Argus*

Imperial War Graves Commission

> Lieutenant Colonel C.E. Hughes, CBE, with Australian and New Zealand officers and NCOs of the commission — works to improve piers, secure a water supply and establish wireless stations

Allied troops *(until 22 September)*

> French detachment
>
>> One company infantry (7th Company, 66th Regiment)
>>
>> One troop cavalry (1st Moroccan Spahis)
>
> Italian detachment
>
>> One company infantry (313th Infantry Regiment)

2 OCTOBER 1922

28th Division

General Officer Commanding, Major-General T.O. Marden, CB, CMG

Divisional troops

  Cavalry

    "B" Squadron, 3rd The King's Own Hussars (less one troop)

  Artillery

    17th Brigade RFA

      13th Battery RFA (4 × 18-pdrs)
      26th Battery RFA (4 × 18-pdrs)
      1st RN Battery (6 × 12-pdrs)

    19th Brigade RFA

      29th Battery RFA (6 × 4.5" howitzers)
      96th Battery RFA (6 × 18-pdrs)
      3rd RN Battery (6 × 12-pdrs)
      5th RN Battery (3 × 12-pdrs)

    Pack Brigade RGA

      1st Pack Battery RGA (4 × 3.7" howitzers)
      14th Pack Battery RGA (4 × 3.7" howitzers)

3rd Medium Brigade RGA

    10th Howitzer Battery RGA ($4 \times 6''$ howitzers)

    34th Medium Battery RGA (4×60-pdrs)

1st Composite Battery RGA

    $2 \times 6''$ howitzers and 2×60-pdrs

Royal Engineers

    No. 2 Section, 55th Field Company, RE

    24th (Fortress) Company, RE

Infantry Brigades

    83rd Infantry Brigade

        2nd Battalion, Royal Sussex Regiment

        1st Battalion, The Loyal Regiment

        1st Battalion, The Gordon Highlanders

    85th Infantry Brigade

        1st Battalion, King's Own Scottish Borderers

        2nd Battalion, Sherwood Foresters

        2nd Battalion, Highland Light Infantry

Kelia area

    21st Stationary Hospital

View over Maidos to Kelia Bay. Anchored off Kelia are the aircraft carriers *Ark Royal* (left) and *Argus* (right).

Royal Air Force

Constantinople Wing

RAF DARFOR, Dardanelles

    267 Squadron

        Fairey IIID seaplanes

    203 Squadron *(until 20 December 1922)*

        Gloster Nightjar single-seater fighters

    4 Squadron

        Bristol F2b fighter and reconnaissance aircraft

Royal Navy

Mediterranean Fleet

Dardanelles Force

    4th Battle Squadron, including battleships *Ajax*, *Centurion*, *Benbow*, *Marlborough* and *King George V*

    3rd Light Cruiser Squadron — HM ships at Chanak on 15 October 1922 were *Cardiff*, *Centaur*, *Caradoc* and *Concord*

    7th Destroyer Flotilla: depot ship, 2 flotilla leaders, up to 18 destroyers

    Aircraft carriers: *Argus*, *Ark Royal*, *Pegasus*

Sherwood Foresters at Nagara.

Passport check.

6 FEBRUARY 1923

28th Division, Attached Troops & RAF

General Officer Commanding: Major-General T. O. Marden, CB, CMG

Divisional troops

    Cavalry

        "B" Squadron, 3rd The King's Own Hussars (less one troop)

    Royal Artillery

        17th Brigade RFA

            13th Battery RFA (4×18-pdrs)

            26th Battery RFA (4×18-pdrs)

            92nd Battery RFA ($6 \times 4.5''$ howitzers)

    Royal Engineers

        12th Field Company RE

        Royal Corps of Signals

        Divisional Signal Company

    Infantry

        2nd Battalion, Royal Fusiliers

    Royal Army Service Corps

        Detail 780 Mechanical Transport Company

        Detail 121 Horse Transport Company

        Detail "E" Supply Company

Royal Army Medical Corps
- 83rd Field Ambulance
- 85th Casualty Clearing Station
- 85th Sanitary Section

Royal Army Veterinary Corps
- 7th Mobile Veterinary Section

Attached Troops (Asiatic Shore)

3rd Medium Brigade RGA
- 10th Battery RGA (4 × 6" howitzers)
- 6th Battery RGA (4 × 6" howitzers)
- 2nd Battery RGA (4 × 60-pdrs)

5th Pack Brigade RGA
- 1st Pack Battery
- 14th Pack Battery

Attached Troops (Gallipoli Peninsula)

Royal Artillery (Heavy)

1st Heavy Brigade RGA
- 1st Heavy Battery (4 × 8" howitzers)
- 2nd Heavy Battery (4 × 8" howitzers)
- 3rd Heavy Battery (4 × 8" howitzers)
- 4th Heavy Battery (4 × 8" howitzers)

5th Medium Brigade RGA

    3rd Medium Battery RGA (4 × 6" howitzers)
    18th Medium Battery RGA (4 × 6" howitzers)
    20th Battery RGA (4 × 6" howitzers)
    22nd Battery RGA (4 × 6" howitzers)
    "S" Coast Battery RGA (late Composite Bty)
    Survey Company RA

Infantry Brigades

    83rd Infantry Brigade

        2nd Battalion, Royal Sussex Regiment
        1st Battalion, The Loyal Regiment

    85th Infantry Brigade

        1st Battalion, King's Own Scottish Borderers
        2nd Battalion, Highland Light Infantry
        2nd Battalion, The Rifle Brigade

Royal Air Force

    267 Squadron RAF
    4 Squadron RAF

The Greek school, occupied by the British Army.

ABBREVIATIONS AND ACRONYMS

BPO	British Forces Post Office
Bn	Battalion
CB	Companion of the Order of the Bath
CBE	Commander of the Order of the British Empire
CMG	Companion of the Order of Saint Michael and Saint George
CO	Commanding Officer
DSO	Distinguished Service Order
GBE	Knight Grand Cross of the Most Excellent Order of the British Empire
GMP	Garrison Military Police
IARO	Indian Army Reserve of Officers
KCB	Knight Commander of the Order of the Bath
MC	Military Cross
NAAFI	Navy, Army & Air Force Institute
NCO	Non-Commissioned Officer
RA	Royal Artillery
RAF	Royal Air Force
RASC	Royal Army Service Corps
RE	Royal Engineers
RFA	Royal Field Artillery
RGA	Royal Garrison Artillery
RN	Royal Navy
RSM	Regimental Sergeant Major
YMCA	Young Men's Christian Association

BIBLIOGRAPHY

Archival sources

National Archives of Australia, Second Australian Imperial Force Personnel Dossiers, 1939–1947, series B883

National Archives of Australia, Citizen Military Forces Personnel Dossiers, 1939–1947, series B884

National Archives of Australia, RAAF Officers Personnel files, 1921–1948, series A9300

National Library of Australia, Trove, Newspapers & Gazettes

The National Archives (UK), British Army service records, pension records and medal index cards, series WO 363, WO 364 and WO 372

The National Archives (UK), Constantinople base records 1918–1923, series ADM 137

University of Birmingham, Cadbury Research Library, Special Collections, Archive of the YMCA (Young Men's Christian Association), GB 150 YMCA

Published sources

'R.E. Work at Kilia and Chanak (with Photo, Plan and Map) Part 1', *The Royal Engineers Journal*, September 1923

Churchill, Winston, *The World Crisis, Part IV, The Aftermath* (London: Library of Imperial History, 1974)

Gwynn, Major-General Sir Charles W., *Imperial Policing* (London: Macmillan & Co., 1934)

Halpern, Paul G. (ed.), *The Mediterranean Fleet, 1919–1929* (Surrey: The Navy Records Society, 2011)

Harington, General Sir Charles, *Tim Harington Looks Back* (London: John Murray, 1940)

Walder, David, *The Chanak Affair* (London: Hutchinson & Co., 1969)

Photographs

From the personal collections of the editor and Jim Grundy.

GET YOUR BOOKS, MAGAZINES AND NEWSPAPERS

FROM

DIRMIKIS & SON
ENGLISH BOOKSELLERS

KILIA AND CONSTANTINOPLE

www.ingramcontent.com/pod-product-compliance
Lightning Source LLC
Chambersburg PA
CBHW011151290426
44109CB00025B/2566